Contents

1　Balwen Welsh Mountain
2　Berrichon du Cher
3　Bleu de Maine
4　Boreray
5　Brecknock Hill Cheviot
6　Cambridge
7　Castlemilk Moorit
8　Charmoise
9　Charollais
10　Cheviot Mule
11　Clun Mule
12　Corriedale
13　Cotswold
14　Est à Laine Merino

15　Exmoor Mule
16　Friesland
17　Galway
18　Gotland
19　Hebridean
20　Highland Mule
21　Icelandic
22　Ile de France
23　Manx Loaghtan
24　Mule Sheep
25　Norfolk Horn
26　North of England Mule
27　North Ronaldsay
28　Ouessant

29　Portland
30　Rouge de l'Ouest
31　Roussin
32　Scotch Mule
33　Scottish Greyface
34　Soay
35　Texel
36　Vendeen
37　Welsh Mule
38　White Faced Woodland
39　Wiltshire Horn
40　Zwartbles

Acknowledgements

So many people have been generous with their help, time and advice that it would be difficult to name everyone and I would risk offending those not mentioned. So thank you to all the breeding societies and associations, individual breeders and farmers without whose help I could not have completed this book.

If there are mistakes they are mine and mine alone.

Picture Credits

Plate (1) Baylham House Farm, (2) Pam and Kelly Hall, (3) Rumwell Bleu du Maine, (4) Stow Estate Trust, (5) Brecknock Hill Cheviot Sheep Society, (6) Cambridge Sheep Society, (7) Baylham House Farm, (8) Genetique Ovine et Developpement, (9) The British Charollais Sheep Society, (10) Mark Lelli DVM, (11) Clun Forest Sheep Breeders Association, (12) Olwen Veevers, (13) The Cotswold Sheep Society, (14) J S Barber, (15) Exmoor Horn Sheep Breeders Society, (16) John Stott, (17) Louise Byrne, (18) British Gotland Sheep Society, (19) Stow Estate Trust, (20) Highlands and Island Sheep Health Association, (21) British Icelandic Sheep Breeders Group, (22) Upra Ile de France, (23) Cathy Wainwright www.manxloaghtansheep.org, (25) Ben and Tory Lugsden, (26) The North of England Mule Sheep Association, (27) The North Ronaldsay Sheep Fellowship, (28) Louise Weeks, (29) Bill Cove, (30) Rouge Sheep Society, (31) Andrea Molyneux, Roussin Sheep Breeder, (32) Catherine A MacGregor www.macgregorphotography.com, (33) The Society of Border Leicester Sheep Breeders, (34) South Oregon Soay Farms, (35) British Texel Sheep Society, (36) Vendeen Sheep Society Ltd, (37) Welsh Mule Sheep Breeders Society, (38) Russell Ashfield, (39) Esmond and Ann Black, Staddlebridge flock, (40) Redgate Zwartbles flock.

Know More Sheep

Jack Byard

First published 2009, reprinted 2010, 2011

Copyright © Jack Byard, 2009

ISBN 978-1-906853-00-6

Published by
Old Pond Publishing Ltd
Dencora Business Centre
36 White House Road
Ipswich IP1 5LT
United Kingdom

www.oldpond.com

Book design by Liz Whatling
Printed and bound in China

Foreword

Here we are again at the beginning of a second book dedicated to the sheep breeds living on Britain's farms today. It seems impossible that a year has gone by since the first book was published but since then I have continued to research these fascinating animals. Some of the breeds have a sad story to tell.

The Boreray, the North Ronaldsay and the Soay are just three of the breeds that now stand on the brink of extinction. Many of these breeds shared the land with the Viking invaders, but like so many things they have become victims of our instant society.

These older breeds grow and mature more slowly than commercial breeds but good food, like good wine, takes time. The keepers of these rare, beautiful animals are prepared to wait. If you have the chance to buy produce from these rare breeds please put your hand into your pocket, by doing this you are helping to preserve an important part of our national heritage.

Walking in the countryside creates a low and possibly non-existent carbon footprint. You are not releasing noxious fumes into the clean countryside air and yes, it is good for you - but don't let that put you off. Breathe the clean air, with maybe a hint of manure.

Please enjoy your country walks and I hope *'Know More Sheep'* will add to your pleasure.

JACK BYARD
Bradford, 2009

Balwen Welsh Mountain

Native to
The Tywi Valley in Wales

Now found
Throughout the British Isles

Protection category

Description

A black, brown or dark grey fleece. It has a white blaze from the top of the head down the nose, four white socks and a half-white tail. The ram has round, low-set horns. The ewe is polled (without horns).

Balwen is Welsh for 'white blaze'. This small but hardy breed has developed through breeding and selection over hundreds of years. They have far fewer health problems than today's larger breeds and are easy to manage; the rattle of a feed bucket and a little patient training are more effective than the traditional sheep dog.

When much of the Balwen's native grazing land was planted with coniferous trees, the population declined. The severe winter of 1947 almost put the Balwen into the history books and it is thought that only one breeding ram remained. Their survival is due entirely to the enthusiasm of local farmers in the 1950s and 1960s who steadily increased numbers. The wool is used for hand spinning and knitwear and the Balwen also produces superb quality meat.

Berrichon du Cher

Native to
Berry at the foothills of the central mountains in France

Now found
Throughout Europe, the British Isles and most major continents

A dense white fleece and clean white face. Both ewes and rams are polled.

In the mid to late 18th century, a passing flock of Merino sheep was used to improve the quality and fine texture of Berrichon du Cher wool. This improvement still exits today. As the demand for wool decreased in the mid 19th century, the meat quality of the breed was improved by a successful cross with the Dishley Leicester. The wool is still in demand by spinners because of the evenness of the fleece, the length of a lock of shorn wool and the strength of staple.

In the year 2000 the Willowfield flock of Berrichon du Cher won the 'Wool on the Hoof' at the Great Yorkshire Show, so the following year the entire clip of all Berrichon was exported to Japan for use in the production of Futons. The same pedigree flock is now used in producing pure wool socks. The Berrichon du Cher also produces high-quality meat.

Description

Bleu du Maine

Native to
The Maine area of France

Now found
Throughout Europe, the British Isles and most major continents

The fleece is cream and very occasionally brown or black. The face, head and legs are grey-blue and free of wool. Often the skin over the entire body is blue, which is the sign of a very good sheep.

The Bleu du Maine is the result of a cross in the 19th century between the Wensleydale, the Leicester Longwool and the now extinct Choltais. The blue colour in the skin comes from the Wensleydale sheep. The Bleu du Maine is noted for giving birth to large numbers of small hardy lambs, who are standing and feeding within minutes of birth and as a result grow quickly. Within eight weeks a lamb can weigh up to 35 kgs on purely the mother's milk and a bit of hay. In the next 14 weeks the ram lambs can weigh up to 65 kgs and the ewes a little less.

The Bleu du Maine gives us superb quality meat and wool which is used mainly for carpets (as is 70% of British wool). Carpet manufacturers want cream wools which can be dyed in a range of colours as well as tough fibres to produce a strong finished yarn.

Boreray

Native to
The island of Boreray on the Scottish west coast

Now found
Throughout the British Isles

Protection category

Description

The face and legs have predominantly black, grey or tan speckles on a white background, frequently forming a collar around the neck. The wool is usually light tan or cream but occasionally grey or dark brown. Both ewes and rams are horned. The ram's horns are large and spiraled

The Boreray is a small primitive sheep about half the size of modern commercial breeds and, unlike commercial sheep, does not require shearing as it sheds its wool naturally in July. The breed was developed in the late 19th century from the Scottish Blackface and an early breed of Hebridean.

In the 1970s, scientists checking the flock on the island of Hirta brought seven sheep back to the mainland. One died immediately and all Boreray in the British Isles are descended from the remaining six. The biggest herd still lives on the island of Boreray as a feral flock. The wool is used for tweeds and carpets and the horn for shepherds' crooks and walking sticks. The Boreray produces fine-quality meat.

Brecknock Hill Cheviot

Native to
The Breconshire area of Mid-Wales

Now found
Throughout Wales

Description

The purebred Brecknock Hill Cheviot can be any colour apart from spotted. The legs and face are white and the ears are erect. They have no wool on the face or legs below the knees. The rams occasionally have horns but the ewes are polled.

This small sheep, which originated in Wales in the mid 17th century, is descended from the original Cheviots and the early Welsh Mountain sheep. The Brecknock Hill Cheviot went unrecognized until the mid 19th century when it was further improved by the Leicester breed. These developments were necessary to enhance the quality of the wool. The Brecknock Hill Cheviot has retained the ability to survive in the harsh hill and mountain conditions of its Scottish Cheviot ancestors. The wool is in great demand for hand spinning and is used for tweeds and knitwear. The Brecknock Hill Cheviot also produces quality meat.

Cambridge

Native to
Cambridge in East Anglia

Now found
Throughout the British Isles

Description

The Cambridge is a medium sized sheep with a white fleece and dark face. The ewes and rams are polled.

Professor John Owen and Alun Davies at Cambridge University founded the Cambridge breed in 1964. The female foundation stock was chosen from the most prolific native breeds, mainly the Clun Forest, Llanwenog *(Thlanwenog)* and Lleyn *(Kleen)*, other breeds that contributed were the Kerry Hill, Ryeland, Border Leicester and Suffolk. Ewes were required to have had three sets of triplets in order to be selected and were crossed with imported Finnish rams.

The particular value of the Cambridge is for crossing with other breeds to produce superior commercial ewes. When crossed with the Texel, in particular, they produce superior quality meat as well as being reproductively efficient. The breed is long-lived and hardy. The wool is used for hand knitwear and hosiery.

Castlemilk Moorit

Native to
The Castlemilk Estate in Dumfriesshire, Scotland

Now found
Throughout the British Isles

Protection category

Description

A light tan or reddish-brown fleece but dark brown next to the skin with a white belly and rump. White is also found around the eyes, lower legs and tail. The ewe's horns curve backwards and outwards; the rams have impressive, heavy, spiraled horns.

The Castlemilk Moorit was developed over a century ago from Moorit (Gaelic for brown), Shetland, Manx Loaghtan and wild Mouflon sheep by the Buchanan-Jardine family to grace the parkland of their home near Lockerbie in Dumfriesshire. In 1970 the original flock was dispersed and the two surviving groups, comprising ten ewes and two rams, are the ancestors of all modern day Castlemilk Moorits.

The Castlemilk's ancestors are ancient breeds each with over a thousand years of history. As a result the Castlemilk Moorit is a hardy breed able to live outdoors on grass all year with only supplementary hay in an extremely bad winter. The wool, even undyed makes excellent tweed, the quality meat has a gamey flavour.

Charmoise

Native to
The hills of France

Now found
*Throughout Europe
and the British Isles*

A white face and fleece with pink skin. The Charmoise is polled.

This is a genuine hill breed, the only top-quality French breed to be classed as a hardy breed or 'race rustique' and the first to be imported into the British Isles from Europe. The Charmoise was created and improved by crossing local French hill sheep with Kent rams.

It is the nature of the Charmoise to forage for food so it is able to survive and thrive on poor pasture. This hardy and disease-resistant animal is able to survive with little attention, making the Charmoise a 'pleasure to shepherd'. The Charmoise provides excellent meat and the wool is used for quality knitwear.

Charollais

Native to
The Saone Loire region of France

Now found
Throughout the British Isles and on most major continents

Description

A large sheep with a fine white fleece. The head has pink skin covered with creamy, sandy or white hair and a distinctive white flash over each eye.

The breed originates in the town of Charolles in the Saone Loire Region of France where it lives alongside the famous Charolais cattle. The breed was developed in the 19th century when the local breeds were crossed with the British Dishley Leicester. The breed has remained pure since that time. The Charollais was imported into the British Isles in 1976 where it is now the second most numerous breed and constantly growing in popularity.

The breed is extremely successful in agricultural shows around the country where it frequently receives the 'Best in Show' award. The meat is of excellent quality and flavour. The majority of the dense, fine quality wool is exported to Japan where it is used in bedrolls.

Cheviot Mule

Native to
The Cheviot Hills in Northumberland

Now found
Throughout the British Isles, Europe and America

Description

The Cheviot Mule has a fine white fleece and blue-tinged face.

The Cheviot Mule is the product of the cross between a Bluefaced Leicester ram and a North Country Cheviot ewe. This popular mule is in constant and increasing demand because it is such a hardy, healthy animal which produces excellent meat from its natural food, grass. The wool is also of superb quality; the coarser wool is used in quality carpet production whilst the finer wool becomes quality knitted products.

Clun Mule

Native to
The Clun Valley of South-West Shropshire and the Welsh border

Now found
Throughout the British Isles, Europe and America

The Clun Mule has a white fleece. The face and legs are nearly always dark and occasionally speckled with white.

This mule is the product of a Bluefaced Leicester ram and a Clun Forest ewe. The Clun Mule is an excellent forager and so produces a wonderfully flavoured meat from its natural food, grass. This is a hardy, healthy animal with a strong resistance to disease. The wool is soft and white having retained the best qualities of the Clun Forest and the Bluefaced Leicester and is in great demand by spinners.

Corriedale

Native to
New Zealand

Now found
Throughout the British Isles, Australia, America and Europe

The face and legs are white but the fleeces range in colour from pale silver to black, various shades of fawn and moorit, red, and occasionally spotted. The Corriedale is polled.

The Corriedale is a fixed Merino x Lincoln Longwool cross, and is the oldest of all crossbred wool breeds. The breed was developed in New Zealand at the end of the 19th century by James Little who arrived from Scotland in 1863. The New Zealand Sheep Breeders Association added the cross to their stock book in 1903 but it was not until 1911 that it became known as the Corriedale.

The Corriedale is a docile, long-lived animal which adapts well to a range of climates including areas of low rainfall. It was initially bred as a dual-purpose animal for food and wool. Their dense fleece, medium-fine with a good length and softness, is in great demand for hand-spun garments, medium-weight outer garments, worsteds and light tweeds. The Corriedale also produces high-quality meat.

Cotswold

Native to
The Cotswolds in the west of England

Now found
Throughout the British Isles, Europe and North America

Protection category

Description

A large sheep. The face is without wool but occasionally there are black spots on the face, legs, ears and hooves. The fleece is white and lustrous with wavy curls and a woolly forelock that falls between the eyes. The ewes and rams are polled.

This ancient and hardy breed has grazed the hills of the Cotswolds since the beginning of the Roman occupation two thousand years ago. The limestone hill pastures were rich in herbs and grasses ensuring a reasonable food supply throughout the year.

During the Middle Ages, wool was a major export and great wealth was created by the Cotswold, which was also known as the Cotswold Lion. 'Wool' churches, including Gloucester Cathedral, were built by the wealthy merchants and are a reminder of the importance of the Cotswold in medieval times. The wool has many uses including carpets, soft furnishings and hand weaving. It is also known as the poor man's mohair. The Cotswold produces a fine-quality meat.

Description

Est à Laine Merino

Native to
Spain and France

Now found
Throughout the British Isles, Europe, and on most major continents

A large sheep with fine white wool. They have no wool around the eyes or below the knees, the head is long and white with drooping ears and both the ewes and rams are polled.

The original Merino was bred for the quality wool which is frequently compared to cashmere. It was developed by a small tribe in North Africa who were defeated by the Spaniards in 1340. Their sheep then fell into the hands of Spanish Royalty and consequently crossed the border into France.

Over two hundred years ago the Merino was used to improve German sheep, primarily in the Alsace Lorraine region of France, these sheep became known as the Est à Laine Merino.

The Est à Laine Merino arrived in the British Isles in 1978. It is a hardy yet docile animal with the ability to survive poor grazing land and the British weather. The breed produces superb quality meat and their fine wool is used for quality knitwear.

Exmoor Mule

Native to
Exmoor

Now found
Throughout the British Isles

Description

A pale beige fleece and a face with a hint of a Roman nose.

The Exmoor mule is the product of a Bluefaced Leicester ram and an Exmoor ewe. The Exmoor Mule is a docile, easily-handled, long-lived and hardy animal with the ability to survive in harsh conditions between sea level and 1500 feet above. The Exmoor mule produces excellent quality meat by eating only its natural food, grass. The quality fleece is used for knitwear and carpets.

Friesland

Native to
Friesland in the north of Holland

Now found
Throughout the British Isles

Description

A pure white fleece and wool-free silky face with pink lips and nostrils. The tail is long and bald and the Friesland is naturally polled.

The Friesland is the only pure dairy breed in the British Isles. The milking of sheep goes back to the very earliest days of agriculture; sheep were milked in Britain for three thousand years before they were replaced by cows around the time of the Black Death. In northern Europe cow's milk has completely taken over, but in southern Europe large numbers of sheep are still milked to produce yoghurt and cheeses such as Feta, Ricotta and Roquefort.

Almost fifty years ago, a few Frieslands were imported into the British Isles to improve the milking and mothering abilities of British sheep and a few pioneers decided to try dairy sheep farming. By the 1980s, it was an established industry producing yoghurt, cheeses and ice cream. Allergic to cow and goat milk? Sheep's milk could be the answer. The Friesland's high-quality wool is used for knitwear and hosiery.

Galway

Native to
Roscommon on the west coast of Ireland

Now found
Throughout Ireland and the British Isles

Description

The Galway has a wavy white fleece with a fine texture. The face is white with dark nostrils and the Galway is naturally polled.

In the 18th century, Dishley Leicester rams were exported to the west coast of Ireland and crossed with the Roscommon to improve the quality of the breed. At the beginning of the 20th century the Roscommon became the Galway. The Galway Sheep Breeders Society was formed in 1923 and, slowly but surely, the breed's popularity increased.

In recent years the Galway has lost ground to more fashionable breeds such as the Suffolk and Texel, so The Galway Society combined with the Irish Genetic Resources Conservation Trust to increase the registered population. The breed is now included in the EC Rural Environment Protection Scheme and receiving a subsidy as an aid to breed conservation. In the British Isles, popularity is growing and the future of the breed is looking bright. The Galway produces superb quality meat and knitting wool.

Gotland

Native to
The island of Gotland off the Swedish east coast

Now found
Throughout the British Isles

Description

A slender, elegant sheep of medium height with a naturally short tail. The Gotland has a black face and legs and a fleece that can be light silver to charcoal grey. Lambs are born black but change to shades of grey within three months.

The breed originated on the island of Gotland, in the Baltic Sea, as a result of the Vikings crossing the native sheep with the Romanov and Karakul sheep brought back from their Russian expeditions. In the 1930s this primitive sheep was subject of a careful breeding and selection process to develop the modern polled Gotland or 'pelt sheep'. Gotlands were first imported into Scotland in 1972 and have now spread to all parts of the British Isles.

The Gotland is hardy, adaptable, friendly and inquisitive. They are mainly kept by smallholders for their quality fleece so there are less than a thousand breeding ewes in Britain. The fleece is used for soft delicate garments and quality-felted products. The pelt is used for coats, hats, slippers and waistcoats, and the Gotland produces excellent flavoured meat.

Hebridean

Native to
The Highlands and Western Islands of Scotland

Now found
Throughout the British Isles

Description

The Hebridean has a black fleece with a black face and legs. The sun can bleach the fleece brown, and some go grey. The horns are heavily ridged. In two-horned animals they sweep up and then back and outwards.

These are descendants of sheep brought to the Western Isles and Highlands of Scotland by the Vikings over a thousand years ago. At the end of the 18th century this small, hardy sheep, able to survive on poor grazing and little supervision, was the main livelihood of shepherds in the area.

The Highland Clearances saw the Hebridean replaced by improved commercial breeds - the Scottish Blackface and the Cheviot - and by the early 20th century the thousand-year history of the Hebridean was obliterated. At about this time the breed appeared in the parklands of country estates in Scotland and England, romantically renamed St. Kilda sheep. These flocks saved the breed from extinction. The wool is popular with hand spinners and knitters. The Hebridean produces high-quality, low-cholesterol meat.

Highland Mule

Native to
Scotland

Now found
Mainly in the Scottish Highlands

Description

A creamy grey fleece and white face with blue-grey speckles.

The product of a Bluefaced Leicester ram and a Blackface or Blackface x Swaledale cross ewe. The lambs are born mainly on Scottish Highland farms which are noted for their extreme weather conditions and few of which have grazing less than 850 feet above sea level. The resultant lambs are vigorous and hardy, a trait inherited from their Blackface mothers and which allows them to survive harsh conditions and grow quickly. If removed from these harsh conditions this Mule mushrooms in size whilst retaining its natural hardiness.

The orange fleece colouring, called bloom, is purely cosmetic for shows or sales and fades within two to three months. The Highland Mule produces quality meat from natural food. The wool is used mainly for knitwear.

Icelandic

Native to
Iceland

Now found
Throughout the British Isles, Germany, the USA and Canada

Description

This small breed's fleece has seventeen distinct colour variations of white, ivory, milky brown, taupe, silver, charcoal, blue-greys, dark brown and black. They can be horned or polled. In their homeland some sheep have four horns but this variation has not yet been seen in Britain.

The Vikings took the breed to Iceland between 870 and 930AD. As it has remained genetically unchanged for 1100 years it is possibly the oldest domestic purebred in the world. The Icelandic was imported into the British Isles in 1979 where it remains uncommon.

The Icelandic is long lived, hardy and intelligent. They are dual-coated, which is typical of primitive breeds. The coarse outer layer, the 'tog', is about 48 cm long and protects against harsh weather; the soft dense inner layer, the 'thel' is about 7 cm long, soft and luxurious. This wool is the origin of the Lopi yarn. The wool is the lightest and warmest in the world, prized by hand spinners and used for baby knits and felted goods. The meat is of the highest quality. The Icelandic is excellent for rough grazing land.

Ile de France

Native to
The Ile de France region surrounding Paris

Now found
Throughout the British Isles, Europe, North America, Canada, Africa and China

Description

A close white fleece with wool on the head, forehead and down to the knees. The lips and nose are pink and the ears are horizontal. Ewes and rams are both polled.

The Ile de France is an integral part of sheep farming in France today. After the wool market slumped at the beginning of the 19th century French sheep were bred for food. In 1824 Professor Auguste Yvart, a professor at Maison-Alfort National Veterinary College decided to develop a more commercial breed, and in 1832 crossed the English Dishley Leicester (the modern Leicester Longwool) with the Rambouillet Merino and later the Mauchamp Merino. Within fifty years he had created a new breed and in 1922 it was named the Ile de France.

The Ile de France arrived in the British Isles in the 1970s. The fine wool is used for hosiery, knitwear and quality textiles. The sheep produces high-quality meat.

23

Manx Loaghtan

Native to
*The Isle of Man,
off the west coast
of England*

Now found
*Throughout the British
Isles and Europe*

Protection category

Description

The Manx Loaghtan has chocolate-brown wool. The face and legs are without wool and are also brown. Newborn lambs are black and start to turn brown at about two weeks old. The ewes and rams have two, four or more horns; the horns on the ewes are smaller.

The Manx Loaghtan (pronounced *Locktun*) is a small, primitive, rare breed. The Loaghtan has been grazing the Manx hills for over a thousand years. Some think the Loaghtan is a native to the Isle of Man, and others believe that it originates from a prehistoric breed in the isolated areas of North-West Europe.

The breed is a slow-maturing animal which survives well on a poor diet. The Loaghtan will sometimes shed its wool in the spring. The wool, which spins into a light warm chocolate colour, is used for Manx tartans, hosiery, knitted garments and woven into lightweight fabrics. The Loaghtan produces lean, low-cholesterol, full-flavoured meat.

24

Mule Sheep

Mule sheep on the facing pages:

(1) Highland
(2) Welsh
(3) Clun
(4) Cheviot
(Centre) Bluefaced Leicester

Description

The Mule, a commercial type of sheep, is the first-generation cross between two breeds. Bluefaced Leicester rams (shown in the companion book *'Know Your Sheep'*) sire the majority of Mules in the British Isles. The dams will be ewes of one of the purebred upland breeds, selected for their ability to thrive best in the locality.

The Mule has the strength, longevity and hardiness of the upland ewe and the excellent meat and wool qualities of the ram. The result is the best of both worlds. The quality wholesome meat from these sheep, fed on grass, is exactly what the health conscious are demanding. The wool is highly prized by home spinners and is used mainly in the manufacture of knitwear and carpets.

Of course, the Mule sheep must not be confused with the other mule – the cross between a horse and a donkey.

25

Norfolk Horn

Native to
Norfolk

Now found
Throughout the British Isles

Protection category

Description

A medium sized sheep with a white fleece. The face and legs should be black or brown and free from wool. Sometimes the sheep will have a black or brown chest and belly. The lambs are born mottled and turn white with age. Both rams and ewes have spiraled horns.

A very handsome and ancient breed. For over three hundred years the Norfolk horn evolved on the sandy heathlands of Breckland in Norfolk. On the south-eastern coast, the country is rugged, food is sparse, and of poor quality; it was in this cold, dry windy area that the Norfolk Horn adapted and developed into the breed it is today.

Due to changing tastes, by 1919 only one flock remained in the entire British Isles, so in 1923 the Norfolk Horn Breed Society was formed. The wool is in demand by hand spinners. The Norfolk Horn produces a wonderful flavoured meat.

26

North of England Mule

Native to
The North of England

Now found
Throughout the British Isles and on most major continents

Description

A pale beige or biscuit colour with a brown-and-white face, free of wool and with the hint of a Roman nose. The ears and legs are white with brown markings. The fleece is long and lustrous.

I cannot improve on this quote from the North of England Mule Association.

'The increasing interest in the North of England Mule has assured its place in every parish in the country. This medium-sized crossbred sheep, sired by the Bluefaced Leicester, has a Swaledale or Northumberland-type Blackface dam. The latter two breeds born and reared on the northern fells and moors are noted for qualities of hardiness, thriftiness and longevity.'

The first wool from a North of England Mule hogg is superb and is used for carpets and by hand spinners who appreciate the long, crimped staple.

27

North Ronaldsay

Native to
North Ronaldsay, one of the Orkney Islands.

Now found
Throughout the British Isles

Protection category

Description

A small, fine-boned animal with a very small head and wool of almost any colour. Ewes have a dished face and can be either horned or polled. The rams have heavy horns.

A seaweed-eating Orkney sheep. The breed has a significant history: in Skara Brae on mainland Orkney, remains were found of what was thought to be a North Ronaldsay type of sheep dating back to the Bronze Age. The breed has changed very little in this time with breeders resisting temptation to 'improve' it.

On their native island, the North Ronaldsay survives on seaweed, which conveniently acts as a copper blocker as they are highly susceptible to copper poisoning. The wool is much sought after by textile designers and for use in handicrafts so a small processing plant on North Ronaldsay produces fine-quality yarns. The meat is delicious and is in demand by fine restaurants around the world.

28

Ouessant

Native to
Ouessant, an island off the French coast

Now found
Throughout Europe

Description

The Ouessant is a small sheep. The fleece is black, brown or white and any shade in between; however, the colour is uniform all over the fleece. The ewes are polled but occasionally have horn knobbles and the rams have impressive curved horns.

The Ouessant is the smallest sheep in the world; the tallest are 50 cm. Until the early part of the 20th century they existed only on the island of Ouessant. It is suggested that the Ouessant are descended from a Viking breed abandoned after an island raid. The size is a result of the harsh living conditions on the island. The Ouessant was saved from extinction by the French nobility who used them as lawnmowers around their elegant chateaux; these lightweight sheep with their delicate hooves did not damage the turf.

The Ouessant is affectionate and easy to keep, requiring minimal land to graze. The wool is not of great quality so is known as felting wool and for making sweaters for fishermen.

29

Portland

Native to
The Isle of Portland off the south-west coast of Britain

Now found
Throughout the British Isles

Protection category

Description

The Portland has a creamy white fleece with tan face and legs and a black nose. Lambs are foxy red turning to creamy white in the first few months. The adults have curved horns, with the ram's horns being heavily spiraled.

The Portland is small compared to commercial breeds and has grazed its native heathland for several hundred years. Once a common sight in Dorset, it is now one of the rarest breeds in England and in 1970 was facing extinction. The Rare Breeds Survival Trust went into action and by 1996 the Portland was no longer at risk of extinction.

The Portland is co-operative if handled frequently, but has a tendency to scatter under the attention of a traditional sheep dog. The Portland is unusual in being able to breed at any time of the year, normally giving birth to only one lamb. The high-quality wool is much valued by hand spinners and the fully-grown rams horns are used by traditional walking-stick makers. The Portland also produces high quality meat which was much praised by George III.

30

Rouge de l'Ouest

Native to
The Loire Valley in France

Now found
Throughout the British Isles and on most major continents

A white fleece, the head and legs are reddish pink and free from wool. Ewes and rams are polled.

The Rouge de l'Ouest means 'The Red of the West', a name that describes its appearance and its origins. It is usually called simply 'The Rouge'. This strong, powerful sheep has a short, fine and dense fleece which gives excellent protection against the harshest of weather conditions.

The Rouge was primarily a dairy sheep, with its thick rich milk being used to produce Camembert cheese. It is now raised primarily for meat. The French breeders concentrated on making The Rouge a dual-purpose breed and this was one of the main reasons that the breed was imported into the British Isles. The Rouge Society was formed in 1986.

31

Roussin

Native to
The La Hague region of France

Now found
Throughout the British Isles and on most major continents

Description

A white fleece and a brown head and legs which are free from wool. Both ewes and rams are polled.

The Roussin is called 'The Roussin de la Hague' in France. In the 18th century, Brittany Heath sheep lived on the moors and dunes of the coastal area in Northern France. In the early 20th century, these were crossed with the Dishley Leicester and the Southdown to improve their size and quality. In 1960, the breed was further improved by the use of the Suffolk and the Avranchin breeds. The Avranchin is a grassland breed adapted to harsh coastal conditions. The Roussin breed development was closed in 1977 and the characteristics were set.

Coming from a coastal region, the Roussin is a hardy adaptable animal with the ability to survive in areas of strong winds, high rainfall and poor-quality grazing. The Roussin produces superb meat. The wool is used for high-quality fabrics and is in great demand by hand spinners.

32

Scotch Mule

Native to
Scotland

Now found
Throughout the British Isles

Description

A medium-sized sheep with a fine white curled fleece and a brown-mottled face. The Scotch Mule is polled.

The Scotch Mule is a cross between a Bluefaced Leicester ram and a Blackface ewe which has been bred in large numbers since the early 1900s. A hardy, disease resistant animal, the Scotch Mule is adaptable to most farming conditions from the harsh uplands to the improved forage and weather conditions of the lowlands. It produces quality meat from natural unadulterated grass and the wool is used mainly for knitwear and tweeds.

The photograph shows a Blackface ewe and her Scotch Mule lambs

33

Scottish Greyface

Native to
Scotland

Now found
Throughout the British Isles

Description

A large sheep with a thick, long, white fleece and a brown-speckled grey face. The Scottish Greyface is polled.

The Scottish Greyface is a cross between a Border Leicester ram and a Scottish Blackface ewe. As with all Mules the cross produces the finest qualities from both breeds: fine quality meat and strong, hard-wearing wool which is mainly used for knitwear and in the carpet industry.

To see a Scottish Greyface on the hillside you could be forgiven for thinking the fleece is grey. The fleece appears grey because of lanolin, natural wool fat, which helps keep the fleece waterproof but also attracts the dirt. Lanolin is present in all-wool bearing animals. Commercially, lanolin is used as a rust preventer, a lubricant and in cosmetics; medical grade lanolin is used in ointments to soothe the skin.

34

Soay

Native to
The island of Soay off the west coast of Scotland

Now found
Throughout the world, especially in the USA and Canada

Protection category

Description

The Soay has a chocolate or fawn fleece. The face and legs are brown or tan with light markings over the eyes, muzzle and lower jaw. There is no wool on the face or legs. The rams have strong-ridged horns and the ewes can be horned or polled.

It is believed the sheep arrived with the first human inhabitants four thousand years ago and were well established when the Vikings arrived in the 9th and 10th centuries. The Soay is possibly one of the most primitive sheep breeds in the world, remaining unchanged for thousands of years. The breed is very similar to the feral Mouflon grazing the hills of Corsica, Cyprus and Sardinia. Until the early 20th century, the breed was only to be found on the island of Soay but in 1932 a flock of 107 were rounded up and brought to the main island of Hirta.

The Soay shed their wool naturally each year and it is used for hand spinning and hand-knitting wool. The meat is delicious and is in demand by the finest restaurants around the world.

35

Texel

Native to
The island of Texel off the north-west coast of Holland

Now found
Throughout the British Isles, Europe, America, Australia and New Zealand

Description

A grey highly crinkled fleece, fine white hair on the head and the occasional black spot on the ears. The ears are carried at ten-to-two.

The Texel has been in existence since Roman times and has, over hundreds of years, been improved by selective breeding and, latterly, the introduction of British breeds. The Lincoln, Border Leicester, Leicester, Southdown, Wensleydale and Hampshire Down have added their genes to improve the breed.

The Texel was imported into the British Isles in 1970 and has become a firmly established breed. There are more Texel sheep in the world than any other breed and it has the largest sheep society in Europe (apparently). The Texel produces excellent meat and the wool is used for carpets and duvets.

36

Vendeen

Native to
The Vendée region of France

Now found
Throughout the British Isles and Europe

Description

The dense white fleece covers the whole body to the knees. The lower legs and face are covered with dark brown to grey hair. The ewes and rams are polled.

The Vendeen has been bred for over five hundred years in the Vendée area of France and folklore tells us it was interbred with sheep rescued from the wrecks of the Spanish Armada in 1588. The breed was improved by using Southdown rams imported in the 19th century.

The Vendeen was first imported into the British Isles in 1981 to improve the birth rate and hardiness of commercial breeds and the first ewe lambs imported all gave birth to twins. The British Vendeen Sheep Society was formed in the same year. The wool is used for hand knitting, quality fabrics and flannel. The Vendeen supplies high-quality meat for both the home and French markets.

37

Welsh Mule

Native to
Wales

Now found
Throughout Wales

Description

The Welsh Mule has a long, lustrous, crimped fleece. The face colour varies from white to a dark mottled or speckled depending on the breed of the dam.

The product of a Bluefaced Leicester ram and one of the hardy Welsh Mountain breeds: the Welsh Mountain; Welsh Hill Speckled Face or Beulah. The Welsh ewes have a reputation for being hardy, healthy and good at foraging for food while being exceptional mothers. The Welsh Mule produces excellent quality meat. The fleece is very fine and used for high-quality woollen products including cloth.

38

White Faced Woodland

Native to
The Pennines in Yorkshire

Now found
Throughout the British Isles

Protection category

Description

A large sheep with a white fleece, broad white face and a pink or part-pink nose. The legs are white and free of wool. Both ewes and rams have horns; the rams have heavy outward spiraling horns.

The White Faced Woodland originated on the borders of Derbyshire, Cheshire and Yorkshire, and records show that the breed was being sold at sales in the area in 1699. The White Faced Woodland is related to the Lonk and the Swaledale and has inherited its remarkably fine-quality wool from an import of Merino sheep in the 18th century.

The popularity of the White Faced Woodland declined during the First World War and this continued until 1973 when the Rare Breeds Survival Trust was formed and the breed was saved from extinction. The wool is used for hand-knitting wool, blankets and carpets. The White Faced Woodland also produces fine-quality meat.

39

Wiltshire Horn

Native to
The Wiltshire Downs in southern England

Now found
Throughout the British Isles, Australia, New Zealand, the West Indies, South America and Africa

Description

This large sheep has long legs and is white all over. They have a Roman nose and ears which are horizontal to the head. The ewes have small horns and the rams' are heavy and curled.

This sheep is almost without wool, making it an ideal export to countries nearer the equator. The wool is 25-30 mm long and not sheared but self-shed by the sheep rubbing up against fences and hedges. The wool then degrades. During the 18th century it is estimated that up to seven hundred thousand Wiltshire Horns grazed the Wiltshire Downs and pastures. It was clearly the dominant breed of the area, answering the particular needs of the district, including manure for the fields, food and some wool for clothing.

In the late 18th century the Wiltshire Horn began to decline in number, giving way to other improved breeds. In recent years it has been in demand both for its superior quality meat and for its absence of fleece. Wool is no longer profitable and farmers appreciate the cost advantages of not having to shear the flock.

40

Zwartbles

Native to
*Friesland in the
north of Holland*

Now found
*Throughout the
British Isles and
Europe*

Description

The fleece ranges in brown from almost black to chocolate. The head is black with a white blaze from the top of the head to the nose. They also have a white tip to the tail and white socks on the rear legs. White socks on the front legs as well are considered ideal. The Zwartbles is polled.

Zwartbles means 'black with a white blaze'. Dutch farmers in the province of Freisland bred the Zwartbles early in the last century but as farming priorities changed they were no longer profitable, and were saved from extinction by the Dutch Rare Breeds Survival Trust. They were introduced into the British Isles in the 1980s, have grown steadily in popularity, and will graze contentedly in the lowland and mid-altitude conditions of England and Wales.

The Zwartbles are docile, friendly sheep and, being naturally tame, prefer to be led rather than herded. The wool is sought after by spinners and weavers and is increasingly popular with felt makers. The meat is very lean and sweet.

Sheep Talk

There are many words associated with sheep which, to those not involved in farming, are a foreign language. Here are just a few of those terms:

Ewe	A female sheep over a year old
Ram	A male sheep over a year old
Dam	A sheep's mother
Sire	A sheep's father
Yearling	A one-year-old animal
Hogg	A yearling sheep, not yet shorn
Lamb	A young unweaned sheep
Mule	A cross between a Bluefaced Leicester and a purebred upland ewe
Dual-purpose	An animal bred to produce more than one product
Pelt	An animal's skin with the fur left on
Staple	Unmanufactured wool
Worsted	A fine wool fabric
Polled	Without horns
Roman nose	A nose with a prominent bridge showing a slightly convex nose in profile
Moorit	A shade of pale brown